DOT-VNTSC-FAA-02-08

FogEye UV Sensor System Evaluation: Phase II Report

Kevin L. Clark
David C. Burnham
Leo Jacobs

Research and Special Programs Administration
John A. Volpe National Transportation Systems Center
Cambridge, MA 02142-1093

Final Report
November 2002

Prepared for

Federal Aviation Administration
800 Independence Avenue, SW
Washington, DC 20591

This document is available to the public
through the National Technical Information
Service, Springfield, VA 22161

U.S. Department of Transportation
Federal Aviation Administration

METRIC/ENGLISH CONVERSION FACTORS

ENGLISH TO METRIC

LENGTH (APPROXIMATE)
- 1 inch (in) = 2.5 centimeters (cm)
- 1 foot (ft) = 30 centimeters (cm)
- 1 yard (yd) = 0.9 meter (m)
- 1 mile (mi) = 1.6 kilometers (km)

AREA (APPROXIMATE)
- 1 square inch (sq in, in^2) = 6.5 square centimeters (cm^2)
- 1 square foot (sq ft, ft^2) = 0.09 square meter (m^2)
- 1 square yard (sq yd, yd^2) = 0.8 square meter (m^2)
- 1 square mile (sq mi, mi^2) = 2.6 square kilometers (km^2)
- 1 acre = 0.4 hectare (he) = 4,000 square meters (m^2)

MASS - WEIGHT (APPROXIMATE)
- 1 ounce (oz) = 28 grams (gm)
- 1 pound (lb) = 0.45 kilogram (kg)
- 1 short ton = 2,000 pounds (lb) = 0.9 tonne (t)

VOLUME (APPROXIMATE)
- 1 teaspoon (tsp) = 5 milliliters (ml)
- 1 tablespoon (tbsp) = 15 milliliters (ml)
- 1 fluid ounce (fl oz) = 30 milliliters (ml)
- 1 cup (c) = 0.24 liter (l)
- 1 pint (pt) = 0.47 liter (l)
- 1 quart (qt) = 0.96 liter (l)
- 1 gallon (gal) = 3.8 liters (l)
- 1 cubic foot (cu ft, ft^3) = 0.03 cubic meter (m^3)
- 1 cubic yard (cu yd, yd^3) = 0.76 cubic meter (m^3)

TEMPERATURE (EXACT)
$[(x-32)(5/9)]$ °F = y °C

METRIC TO ENGLISH

LENGTH (APPROXIMATE)
- 1 millimeter (mm) = 0.04 inch (in)
- 1 centimeter (cm) = 0.4 inch (in)
- 1 meter (m) = 3.3 feet (ft)
- 1 meter (m) = 1.1 yards (yd)
- 1 kilometer (km) = 0.6 mile (mi)

AREA (APPROXIMATE)
- 1 square centimeter (cm^2) = 0.16 square inch (sq in, in^2)
- 1 square meter (m^2) = 1.2 square yards (sq yd, yd^2)
- 1 square kilometer (km^2) = 0.4 square mile (sq mi, mi^2)
- 10,000 square meters (m^2) = 1 hectare (ha) = 2.5 acres

MASS - WEIGHT (APPROXIMATE)
- 1 gram (gm) = 0.036 ounce (oz)
- 1 kilogram (kg) = 2.2 pounds (lb)
- 1 tonne (t) = 1,000 kilograms (kg) = 1.1 short tons

VOLUME (APPROXIMATE)
- 1 milliliter (ml) = 0.03 fluid ounce (fl oz)
- 1 liter (l) = 2.1 pints (pt)
- 1 liter (l) = 1.06 quarts (qt)
- 1 liter (l) = 0.26 gallon (gal)
- 1 cubic meter (m^3) = 36 cubic feet (cu ft, ft^3)
- 1 cubic meter (m^3) = 1.3 cubic yards (cu yd, yd^3)

TEMPERATURE (EXACT)
$[(9/5) y + 32]$ °C = x °F

QUICK INCH - CENTIMETER LENGTH CONVERSION

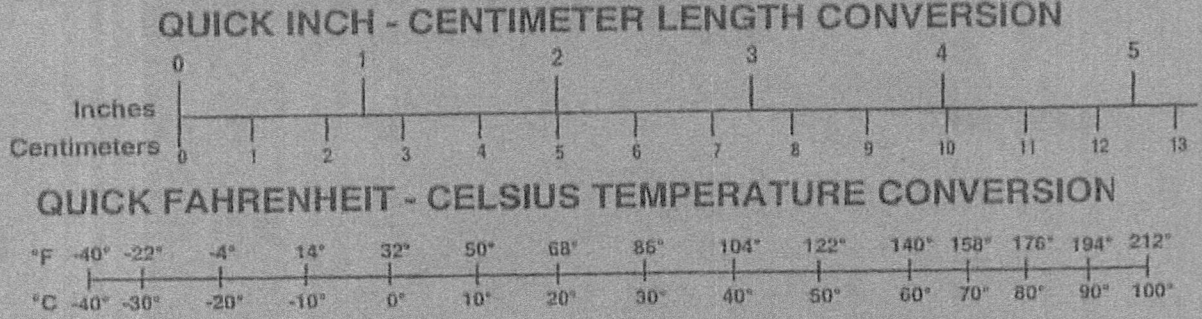

QUICK FAHRENHEIT - CELSIUS TEMPERATURE CONVERSION

°F	-40°	-22°	-4°	14°	32°	50°	68°	86°	104°	122°	140°	158°	176°	194°	212°
°C	-40°	-30°	-20°	-10°	0°	10°	20°	30°	40°	50°	60°	70°	80°	90°	100°

For more exact and or other conversion factors, see NIST Miscellaneous Publication 286, Units of Weights and Measures. Price $2.50 SD Catalog No. C13 10286

TABLE OF CONTENTS

1. INTRODUCTION 5
   1.1 Background 5
   1.2 Test Objective and Methodology 5
   1.3 Operational Applications 5
2. TEST CONFIGURATIONS 7
   2.1 FogEye Transmitter and Receiver Characteristics 7
   2.2 AC-Powered Trip Wires 9
   2.3 Battery Operated Trip Wires 13
3. TEST CHRONOLOGY 14
   3.1 Trip Wires on Scaffold Tower 14
   3.2 Trip Wires on 300-Foot Baseline 14
   3.3 Solar Blind Check 15
   3.4 Airfield Operational Tests 15
4. DATA COLLECTION AND PROCESSING 16
   4.1 Data Recording 16
       4.1.1 Trip-Wire Test (WTF Compound – Test Van) 16
       4.1.2 Trip-Wire Test (Airfield – F-15s and Support Equipment) 20
       4.1.3 Fog Test 23
   4.2 AGC Response 23
5. DATA ANALYSIS 24
   5.1 Beam Breaker Results for AC-Powered Trip Wires 24
   5.2 Beam Breaker Results for Battery-Powered Trip-Wire Units 24
   5.3 Fog Results for AC-Powered Trip-Wires 24
       5.3.1 Blocked Unit 2 24
       5.3.2 Transmissometer Unit 3 26
6. OPERATIONAL CONSIDERATIONS 29
7. CONCLUSIONS 33
8. FUTURE TESTING 34

PREFACE

FogEye technology offers the potential for operation of electro optical sensors and systems that function "hands-off", over extended distances during varying atmospheric conditions, day and night. The technology has been shown to employ solar blind ultraviolet radiation and special electro optical hardware to achieve immunity to natural background radiation. Solar blind ultraviolet radiation may also have favorable atmospheric transmission properties. A basic FogEye hardware complement consists of a Transmitter, a source of ultraviolet radiation, and a Receiver that accepts this radiation and rejects all of the sun's radiation as well as rejecting man-made sources of non solar blind radiation, but, of course, is sensitive to man-made solar-blind radiation.

Congress has requested the FAA to assess this FogEye technology and evaluate the feasibility of applying it to aviation-related problems. The FAA's Office of Surface Technology Assessment (AND-520) has taken responsibility for this investigation and has requested the support of the Volpe National Transportation Systems Center. (Volpe Center)

The FogEye equipment tested has been provided by Norris Electro Optical Systems.

The technical support provided by Norris Electro Optical Systems, specifically Robert Evans and Mike Thorsted, was key to successfully accomplishing the test activities. In addition, this evaluation could not have been successfully completed without the cooperation and help of the 102nd Fighter Squadron of the Massachusetts Air National Guard, stationed at Otis Military Reservation; specifically, Major Mike Dolan, Airfield Operations Officer.

EXECUTIVE SUMMARY

The control of aircraft and vehicles on airport surfaces is changing considerably. With the advent of very low visibility operations, down to 300 feet RVR, a concerted effort to preclude runway incursions has been underway for several years. These efforts include the ASDE-3 Radar and Airport Movement Area Safety System; the ASDE-X Radar and Multilateration Systems; Automatic Dependent Surveillance System Broadcast (ADS-B), and Airport Surface Movement Sensors (AMASS).

The Federal Aviation Administration (FAA) and the commercial aviation industry seek to develop and evaluate technologies that increase safety and efficiency of airport operations under low visibility conditions. The overall goal of these efforts is to provide commercial aircraft with the technology and operating procedures needed for safely achieving the throughput of clear-weather surface operations during adverse-weather conditions. The detection range of systems based on visible light, such as runway lights, is reduced by intervening atmospheric scattering of sunlight during low visibility conditions. FogEye technology operates in the solar-blind ultraviolet region of the spectrum and does not "see" this visible scattering. It may therefore provide detection ranges during low visibility day light conditions that are comparable to nighttime conditions. It may also offer an additional advantage of simplicity and therefore be quite cost-effective, especially when compared to some of the existing efforts.

This report presents the results of the FogEye UV Sensor/System Evaluation for Phase II, which examined trip-wire sensors for detecting aircraft and other vehicles on the airport surface. The test configurations were set up to evaluate the sensor as a runway incursion detection device. The dynamic response of the device was evaluated as well as its performance in fog. The dynamic response tests involved passing a vehicle-mounted object through the beam. This simulated the size and speed of aircraft wheels during "worst case" conditions. Static tests were conducted under similar conditions to determine the impact of forward scattering, due to fog, on the ability of the device to detect the movement of an aircraft's nose wheel. These tests were followed by evaluation in an operational environment. FogEye sensors were placed across a taxiway and their responses to movement of F-15 aircraft observed. In all cases the FogEye trip-wire sensor functioned satisfactorily. Earlier, Phase I tests, validated that the FogEye system was blind to solar radiation and that atmospheric propagation of solar blind radiation was at least comparable to visible propagation, and perhaps more favorable.

The data from this test indicate:

- The sensor performed well as a trip-wire sensor for aircraft and vehicle detection under bright light conditions. Utilizing a static target on a 300-ft baseline, the sensor performed well during periods of dense fog.

- If used with an alarm system, status lights, or voice system, the FogEye Trip-Wire System would be a feasible sensor for runway incursion detection and prevention.

The future application of FogEye trip wires will depend upon issues of cost, ease of installation, reliability, etc. that were not considered in this evaluation. The hardware configurations used for the evaluations will be repackaged prior to operational demonstrations. The repackaged Transmitters and Receivers must have form, fit, and function characteristics compatible with installations that are integral with runway and taxiway edge lights. An integral installation consists of either a Transmitter or a Receiver. The number of hardware failures experienced during the Weather Test Facility (WTF) test program suggests that reliability will have to be addressed in operational units.

1. INTRODUCTION

1.1 Background

FogEye is a commercial name for solar-blind ultraviolet technology used to penetrate fog. The technology can be applied to circumstances requiring navigation or surveillance during low visibility conditions. Phase I of the FogEye evaluation compared[1] the fog attenuation of FogEye beams to the attenuation measured by standard visible-light transmissometers. Phase II examined the feasibility of FogEye trip wire sensors.

The FogEye Aircraft Presence (trip-wire) Sensor (FEAPS) is intended to autonomously detect the proximate presence of aircraft on runways and taxiways. Each FEAPS consists of a Transmitter and a Receiver. They are separated by the width of a taxiway or runway. Airport demonstration models will be individually packaged for installation, integral with existing taxiway or runway light fixtures. Individual sensors or combinations thereof, can function as triggers for indications of presence of aircraft or vehicles. A system incorporating FEAPS thus performs monitoring and control functions similar to those of a traffic policeman at a vehicle intersection. The goal of Phase II is to demonstrate that the FogEye UV trip-wire sensor can be used to reliably detect aircraft and vehicular traffic in fog and under bright daylight conditions.

1.2 Test Objective and Methodology

The objective of Phase II was to evaluate the operational performance of a FogEye trip-wire system for detecting runway incursions. The evaluation examined trip-wire baselines of 75 and 300 feet to cover the expected range of application to taxiways and runways. Placement of the sensors was consistent with current taxi-way lighting installation.

Because of the problems associated with finding fog conditions to conduct a test, the initial evaluation was divided into two portions that test different aspects of trip-wire performance:

1. The time constants and sensitivity of the FogEye trip-wire sensor were first evaluated under high visibility conditions using a 75-foot baseline. Similar baselines were utilized both at the weather test facility and on the airfield.

2. The sensors were then installed on a 300-foot baseline parallel to a visible-light transmissometer to evaluate the influence of fog on tripwire performance. Because forward-scattered light can route the UV around the aircraft tire being detected, one of the two available detectors was blocked by a circular object that simulated an aircraft tire.

These initial performance evaluations were performed at the Volpe WTF. They were followed by operational evaluations at the Otis Air National Guard airfield.

1.3 Operational Applications

Although the purpose of the test is to evaluate the functionality of the FogEye sensor, operational applications will be reviewed and discussed.

---

[1] Clark, K. L., Burnham, D. C., and Jacobs, L., FogEye UV Sensor System Evaluation: Phase I Report," Report No. DOT-VNTSC-FAA-02-04, Volpe National Transportation Systems Center, Cambridge, MA.

Specific applications have the following questions:

On-Off Switch - can the system allow for automatic sequencing of stop bar and taxiway center-line lead on lights after initiation by the switch controller?

Taxiway Light Sequencing – can the system allow for automated sequencing of segmented taxiway lights to give positive direction information to pilots?

Presence/Memory - can the system indicate the presence of an aircraft/vehicle at any given location on the airport? For aircraft/vehicles stationary on a sensor, can the system maintain a presence detection?

Runway Exit – can the system be used to give a positive indication of aircraft/vehicle being clear of the runway?

Flashing PAPI – can the system provide a signal to an existing PAPI that a potential runway incursion is developing or has actually occurred on the runway site of the PAPI? Receipt of the signal would ultimately result in flashing of the PAPI, thereby warning the flight crew of an approaching aircraft.

Runway Status Lights – can the system provide a signal to lights located within the field of view of a flight crew that has positioned an aircraft and is holding for takeoff at the end of a runway? This signal would cause the threshold located lights to alert the flight crew to hold due to the detection of a potential runway incursion that is developing or one that has actually occurred.

Ground Marker – can the system provide a signal to a device that will transmit an audio signal via a local RF link to a flight crew advising them of the taxiway or runway location through which their aircraft is currently moving? Also can this signal be accompanied with an appropriate broadcast time slot that will allow for additional information to be relayed as a function of the location and/or speed of the aircraft?

2. TEST CONFIGURATIONS

2.1 FogEye Transmitter and Receiver Characteristics

2.1.1 FogEye Aircraft Presence Sensor Characteristics

Table 1 lists the characteristics of the FogEye transmitter and receiver. The receiver beam size of 1.5 degrees is much narrower than that used in the Phase I transmissometer.[1] Receiver 1 had a time constant of 11 msec and Receiver 2 had a time constant of 22 msec. The gain transfer curve for the two units are presented in Figure 12.

Table 1. FogEye Aircraft Presence Sensor Characteristics

Function: Senses the presence or movement of an aircraft at a distinct surface location

Description: Each unit consists of a Transmitter and a Receiver, separated by a distance of 75 to 300 feet

Characteristics

Transmitter	P/N 07MF2046001-4, S.N's 1004, 1005, 1014 & 1015
Wavelength	254 nanometers
Output Power	42 Microwatts/(cm)^2 – Steradian
Beam Width	12^0 half width, half power
Excitation Frequency	120 Hz (1004 & 1005); 2000 Hz (1014 & 1015)
Prime Power	115 VAC, 60 Hz, 20 watts (1004 & 1005)
	12 VDC, 20 watts (1014 & 1015)
Receiver	P/N 07MF4-2047001, S/N's 1001, 1003, 1004 & 1005
Wavelength	254 nanometers
Sensitivity	3 x 10^8 amp./watts
Field of View	1.5^0
Dynamic Range	3 x 10^5
UV/Visible Isolation	>10^6
Prime Power	115 VAC, 60 Hz, 1.5 watts (1001 & 1003)
	12 VDC, 1.5 watts (1004 & 1005)
Signal Outputs	
AGC Level	1-3 VDC; Gain Min. to Max.
Detected Signal	5-.01 VDC; Signal Max. to Min.
Threshold Setting	Variable, from nominal 3 VDC to 0.5 VDC
Presence Detection Indication	TTL format: +5 VDC, sensor activated; 0 VDC, presence detected

Table 1 Cont.

Dynamic Response
   AGC Time Constant   20 seconds
   Signal Time Constant   22 milliseconds (S/N 1001)
                                   11 milliseconds (S/N 1003)
                                   4 milliseconds (1004 & 1005)

Volpe Weather Test Facility Phase II Hardware Configuration/Identification Summary

Unit Title/Description	Transmitter	Receiver
Unit 2 Aircraft Presence Sensor; Beam Blocked by Disc;	07MF4-2046001-4 S/N 1004; T2	07MF4-2047001 S/N 1003 Detector S/N 0464H052 "Receiver 1" 11 msec Signal Time Constant
Unit 3 Aircraft Presence Sensor; Beam Attenuated by Fog; "Transmissometer Unit 3"	07MF4-2046001-4 S/N 1005; T3	07MF4-2047001 S/N 1001 Detector S/N 0465H052 "Receiver 2" 22 msec Signal Time Constant
Unit 4 Aircraft Presence Sensor; for Airfield Operational Evaluation	07MF4-2046001-6 S/N 1014, T6 1000 Hz Excitation Frequency Power Supply: 07MF4-2046501-1 S/N 1001; P1 12 VDC; 7 amp.hr.	07MF4-2047001-1 S/N 1004 4 msec Signal Time Constant Power Supply: 07MF4-2047501-1 S/N 1101, P7_ 12 VDC; 21 amp. hr.
Unit 5 Aircraft Presence Sensor; for Airfield Operational Evaluation (Subject of this Phase II Report)	07MF4-2046001-6 S/N 1015; T7 1000 Hz Excitation Frequency Power Supply: 07MF4-2046501-1 S/N 1004; P4 12 VDC; 7 amp. hr.	07MF4-2047001-1 S/N 1005 4 msec Signal Time Constant Power Supply: 07MF4-2047501-1 S/N 1102, P8_ 12 VDC;; 21 amp. hr.

2.1.2 FogEye Interface Description for Units Under Evaluation

An interface block diagram is provided in Figure 1.

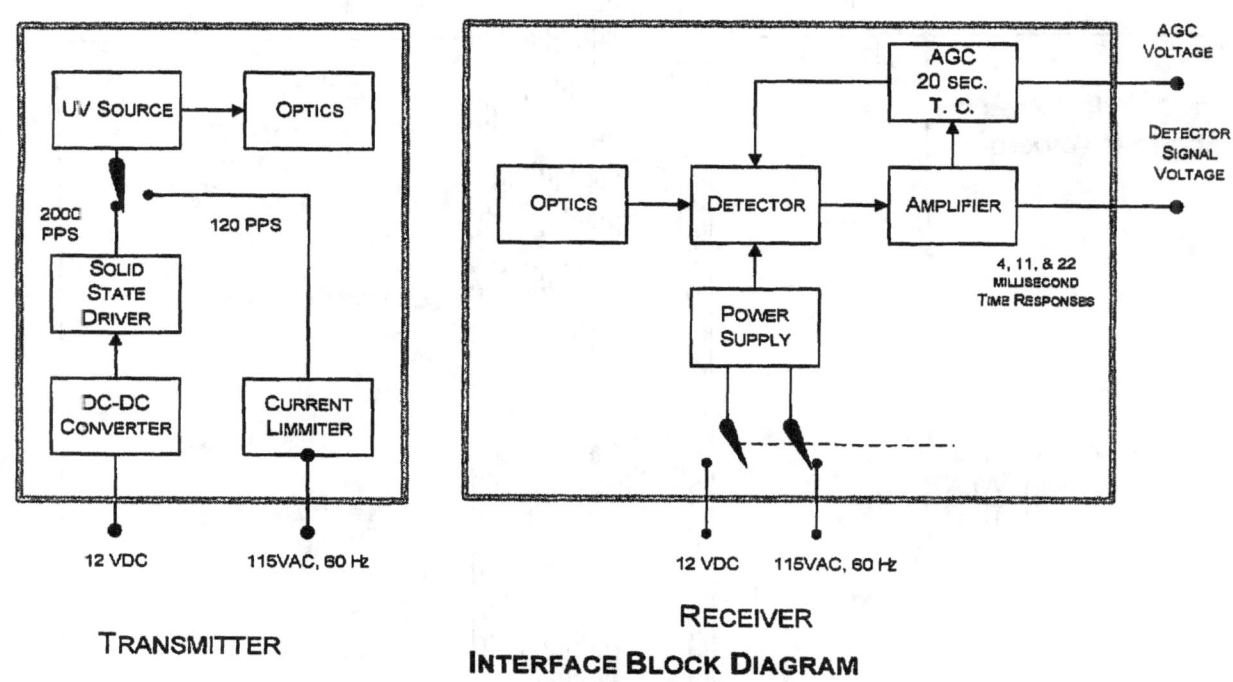

INTERFACE BLOCK DIAGRAM
TEST CONFIGURATION
FOGEYE™ AIRCRAFT PRESENCE SENSOR

Figure 1

2.2 AC-Powered Trip Wires

2.2.1 Dynamic Response

The objective of these tests was to ensure that the trip wire would be sensitive to an interruption caused by movement of an aircraft nose wheel during both the slowest and the fastest set of conditions.

In Phase II two FogEye trip wire sensors were installed on two scaffold towers separated by 75 feet (see Figure 2). The beams could be broken by vertical masks mounted on top of a van that could be driven past the trip wires at speeds up to approximately 30 Mph.

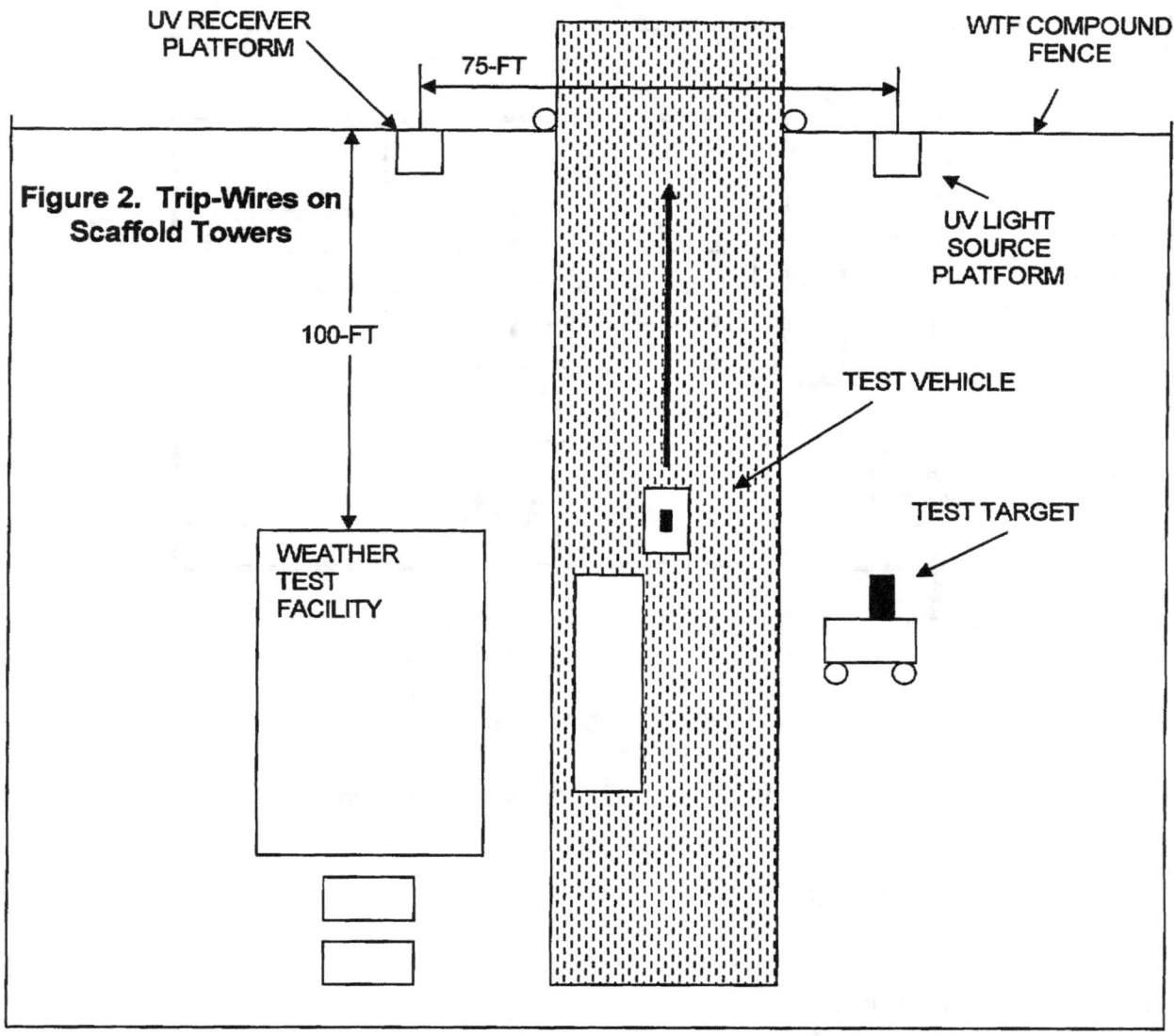

Figure 2. Trip-Wires on Scaffold Towers

Pictures 1 through 4 depict the trip-wire test set-up. Two test platforms were constructed and separated by 75-ft. The first platform was configured with two UV receivers at a height of 9-ft and a Regal laser range finder at a height of 7 ½ -ft. The second platform was configured with a UV Light source at a height of 9-ft.

A van with a 5-inch, 15-inch or 27-inch wide target attached to the roof rack was used as the test vehicle. The 15-inch target simulated the diameter of a typical general aviation (GA) aircraft; the 27 inch target simulated a Boeing 737 nose wheel. The test vehicle was driven past the trip-wire system at speeds ranging from 5 to35 mph. The speed was determined through use of a Regal Laser (Refer to Picture 1). In this configuration, the trip-wire senses the test vehicle's presence by detecting a momentary interruption in a sensor beam that extends across the 75-ft distance between the sensor's

Transmitter and Receiver units. This interruption is caused by interception of the beam by the test target attached to the van. A Campbell Scientific CR-9000 Digital Data Recorder was used to record detection, and response time.

Picture 1. Two UV Receivers and Regal Laser

Picture 2. CR-9000 Digital Data Recorder

Picture 3. UV Light Source

Picture 4. Test Vehicle and 27-inch Target

2.2.2 Performance During Fog

Application of the FogEye technology as a beam breaker could be limited by its own background. That is, a Transmitter's beam could be forward scattered "around" an aircraft's nose wheel during dense fog conditions and hence the Receiver would not detect an interception of the beam due to the scattered radiation. The test was conducted to determine if such a condition does occur and, if so, under what conditions and of what consequence.

After the tests with the van were completed, one UV light source was mounted on the receiver end of the 300-foot visible light transmissometer (see Picture 5) and the two receivers were mounted on the projector tower of the 300-foot visible-light transmissometer (see picture 6). One of the two beams was blocked by a circular stop representing an aircraft tire (15 & 22 inches in diameter) (see Figure 3). The circular stop was aligned by drilling a hole in the center and adjusting the stop location to give the maximum transmission value; the hole was then blocked. (see Picture 7 and 8)

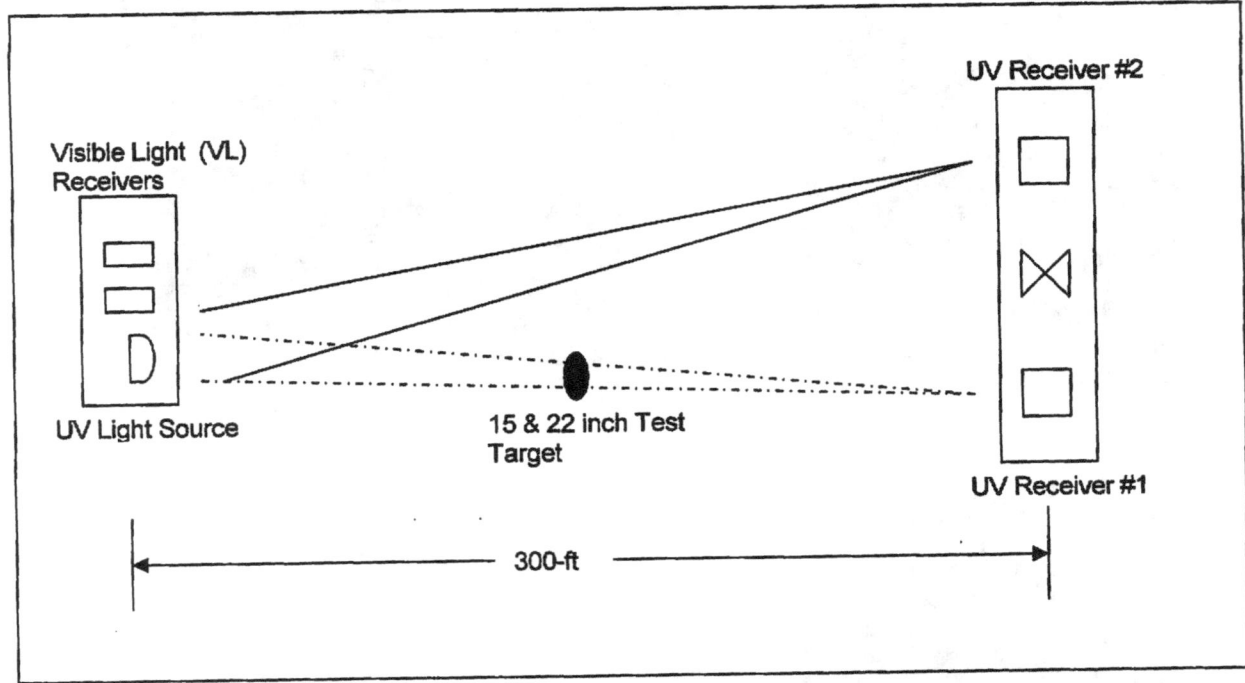

Figure 3. Trip-wires on 300-ft Baseline

Picture 5. UV Light Source

Picture 6. Two UV Receivers

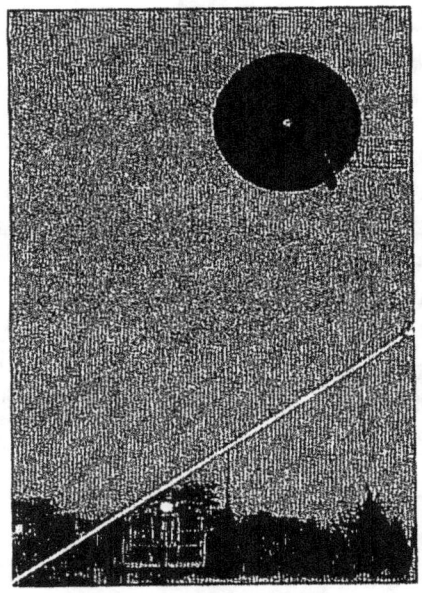
Picture 7. 15-inch Test Target

Picture 8. Test Target and UV Receivers in Background

2.3 Battery Operated Trip Wires

2.3.1 Operational Evaluation

Two additional units, numbers 4 and 5, were supplied for the operational evaluation tests at the Otis airfield. The transmitters and the receivers of both these units were battery powered to allow for ease of deployment on the airfield. In addition, the performance of both these units was improved by circuitry changes that were indicated during tests of the AC powered units. Prior to the operational evaluation, the units were subjected to pre-operational functional tests at the Volpe WTF. The Transmitters, with 1000 Hz UV light sources, and the Receivers, with 4 msec time-constants, were first installed near the ground in the same location shown in Figure 2. The test van was used to block the beams. These tests verified the improved performance expected from the new FogEye trip wires before they were deployed for the operational evaluation phase at the Otis Air National Guard Base. Pictures 9 and 10 depict the setup of the FogEye Sensors on the airfield for the operational evaluation. The distance between the UV light source and receiver was 85-ft. The distance between Sensors 1 and 2 was 25 ft.

Picture 9. Two pair of FogEye Trip Wire Sensors Deployed at F-15 Taxiway

Picture 10. Close-up of F-15 nose-wheel crossing between UV light source and receiver.

3. TEST CHRONOLOGY

3.1 Trip Wires on Scaffold Tower

The trip wire system was installed on 12-ft scaffold towers on a 75-ft baseline on August 8, 2002. Seven calibration runs were conducted with the test vehicle. A 27-inch wide target was utilized to simulate a Boeing 737 size nose wheel. Speeds ranged from 5 to 35mph. Initial run data indicated the sampling rate of the CR-9000, 1000 hz, was effecting the signal oscillations. The sampling rate was reduced to 500 hz. In addition, the time constant on the receivers were changed to 11 msec for receiver 1 and 22 msec for receiver 2. On August 9, 2002, 33 runs were conducted with the test vehicle at speeds of 5 to 30 mph. On August 29, 2002, eight runs were conducted with the test vehicle. Speeds ranged from 5 to 40 mph. A 27-inch wide target was utilized. On September 4, 2002, seven test runs were conducted at speeds of 5 to 30 mph with a 5-inch wide target attached.

3.2 Trip Wires on 300-Foot Baseline

The trip wires were installed on the 300-foot baseline on 9/18/02. The projector was installed next to the visible light receivers (see Picture 5). The two receivers were installed next to the visible light projector (see Picture 6). The 1.5-degree receivers were aligned to give maximum signal (actually minimum AGC voltage). The AGC system tries to keep the signal at 3.0 Volts. The AGC voltage varies from 1.0 Volts to a maximum value of approximately 2.9. After the AGC voltage reaches its upper limit, then lower signals result in lower signal voltages. Figure 4 shows the data on 9/18/02 after the trip-wire Unit 3 was aligned. The first trip-wire unit (11 msec time constant) is termed Unit 2 and the second trip-wire unit (22 msec time constraint) is termed Unit 3.

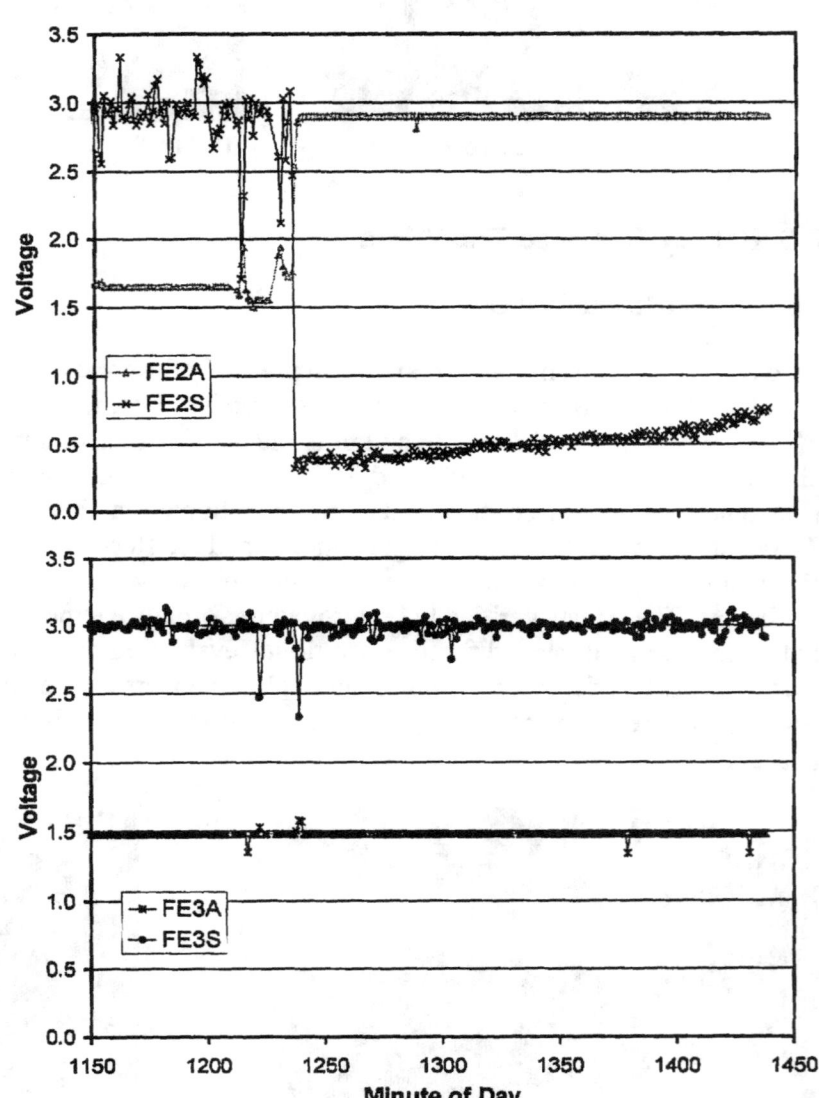

Figure 4. Installation of Trip Wire Sensors on 300-Foot Baseline 9/18/02

Unit 2 (FE2) is the unit that had its beam blocked. Unit 3 is the unit that functioned as a transmissometer. FE2A is the AGC signal for Unit 2 (Receiver 1 with 11 msec time constant) and FE2S is the detector output signal; FE3A is the AGC signal for Unit 3 (Receiver 2 with 22 msec time constant) and FE3S is the detector output signal.

The orientation of trip-wire Receiver 2 was optimized around minute 1220 (top plot of Figure 4); the AGC voltage was about 1.55. The signals from this time were taken as the 100% transmission value for both trip-wire units in the subsequent analysis.

The aircraft tire mask (15" diameter, see Picture 7) with a 2-inch hole in the middle was then centered on the beam around minute 1235; the AGC voltage was slightly higher (about 1.7) than before the blockage. The hole was then blocked; the AGC voltage rose to its limit and the signal voltage dropped to about 0.4 volts. The voltage rose to about 0.7 volts by the end of the day.

3.3 Solar Blind Check

The ac-powered trip-wire receivers were found to be solar blind by turning off the transmitter at noon on 10/1/02. A similar test was conducted with the battery-powered units 4 & 5 on October 7, 2002.

3.4 Airfield Operational Tests

From October 2 – 7 the battery-operated units were calibrated and tested to verify that the modifications introduced corrected the issues of the original units. Following some minor adjustments, each unit was cleared for deployment. Both UV light sources stopped working and had to be reset a number of times the morning of October 8, 2002. Moisture and low temperatures were blamed for the problem. On October 8, 2002, the battery-operated units were installed on a high traffic area of the Otis bravo taxiway. After each day of data-collection the units were removed and re-installed the following day. From October 8 –10 a total of 21 runs were recorded. All targets that crossed the beam were detected and no false alarms or missed detections were registered.

4. DATA COLLECTION AND PROCESSING

4.1 Data Recording

4.1.1 Trip-Wire Test (WTF Compound – Test Van)

The FogEye receivers each provided two signals: Signal Voltage (FEnS) and Automatic Gain Control (AGC) Voltage (FEnA). For the 75-foot trip-wire test they were digitized by a Campbell Scientific Model CR-9000 data acquisition unit, which sampled at a 500 or 1000 Hz rate and recorded the signals to a data file. Data acquisition was initiated just before the van reached the test section. The data could then be plotted for analysis. Figure 5 depicts a CR-9000 plot from August 29, 2002 run. A number of interruptions are shown. The x-axis of the plots is the number of samples.

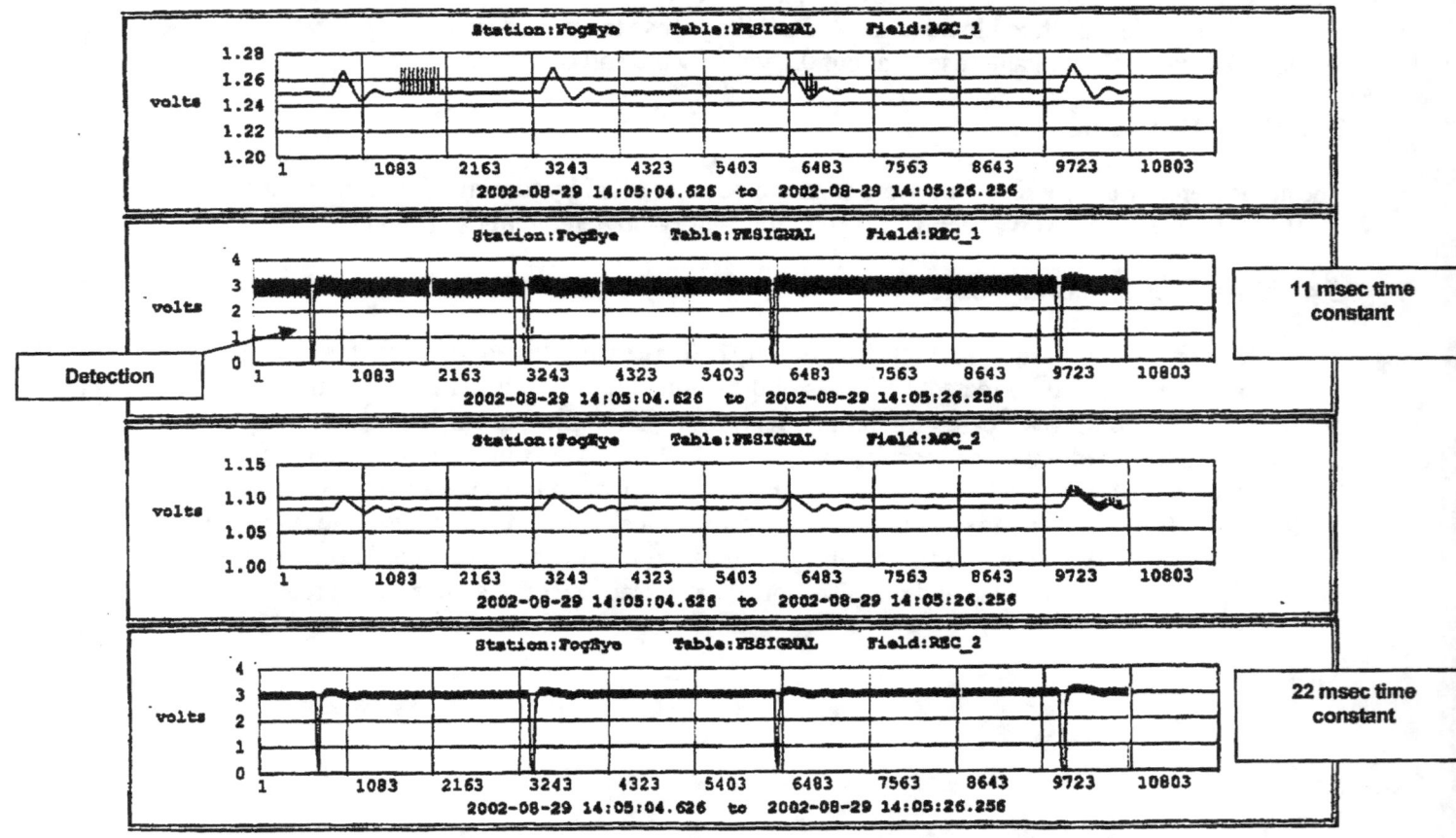

Figure 5. 8/29/02 Run 1 (10 mph) – FESIGN 00

Using an Excel spreadsheet, CR-9000 data were plotted on an expanded scale to detail the trigger and response of each run. Figures 6 - 8 depict several of the plots from August 29, 2002 testing. The plots will be analyzed in the next chapter. *Note: Universal Time (UTC) listed on plots read as minutes, seconds, and 1/10 of a second.*

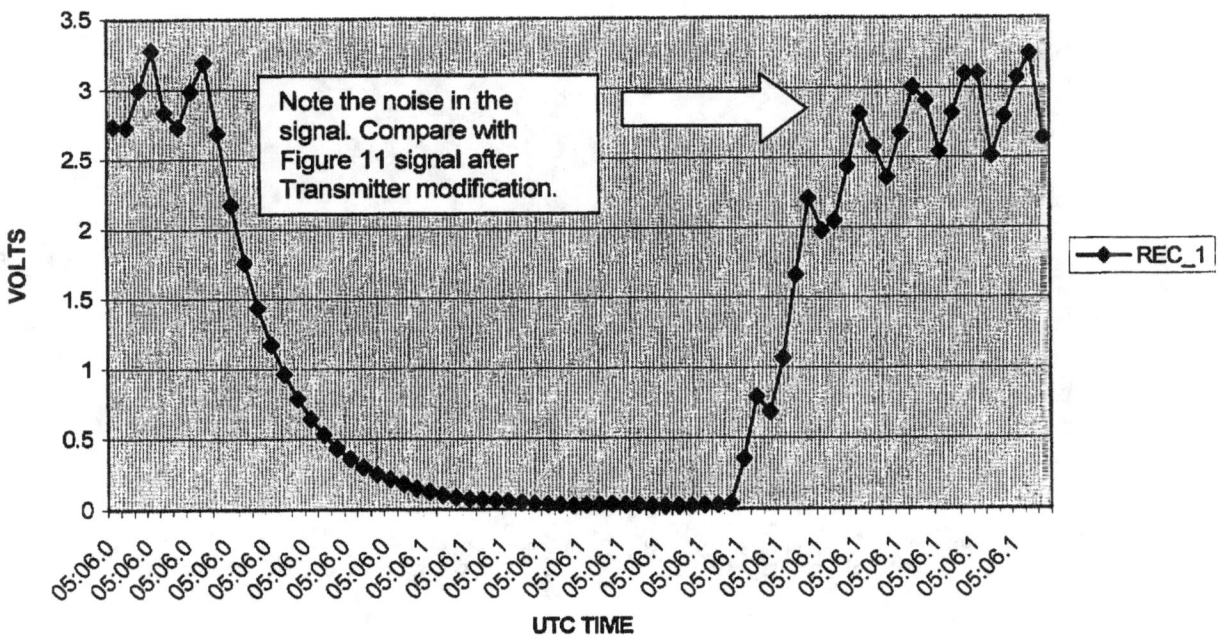
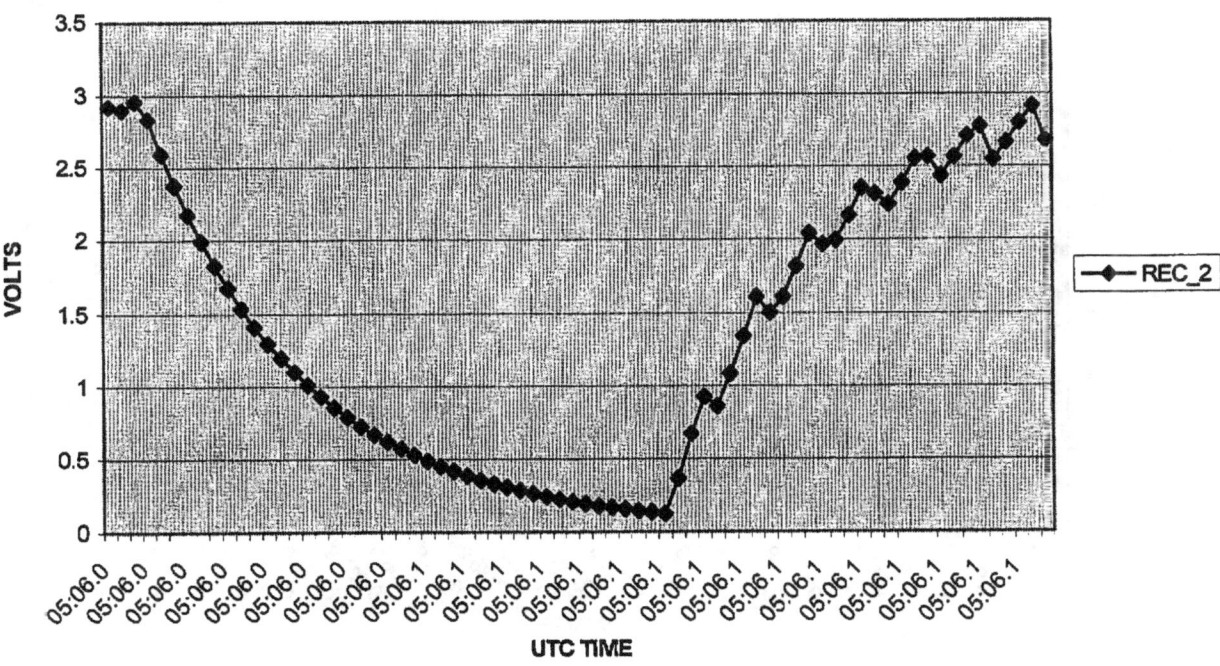

Figure 6. Excel Data Plots

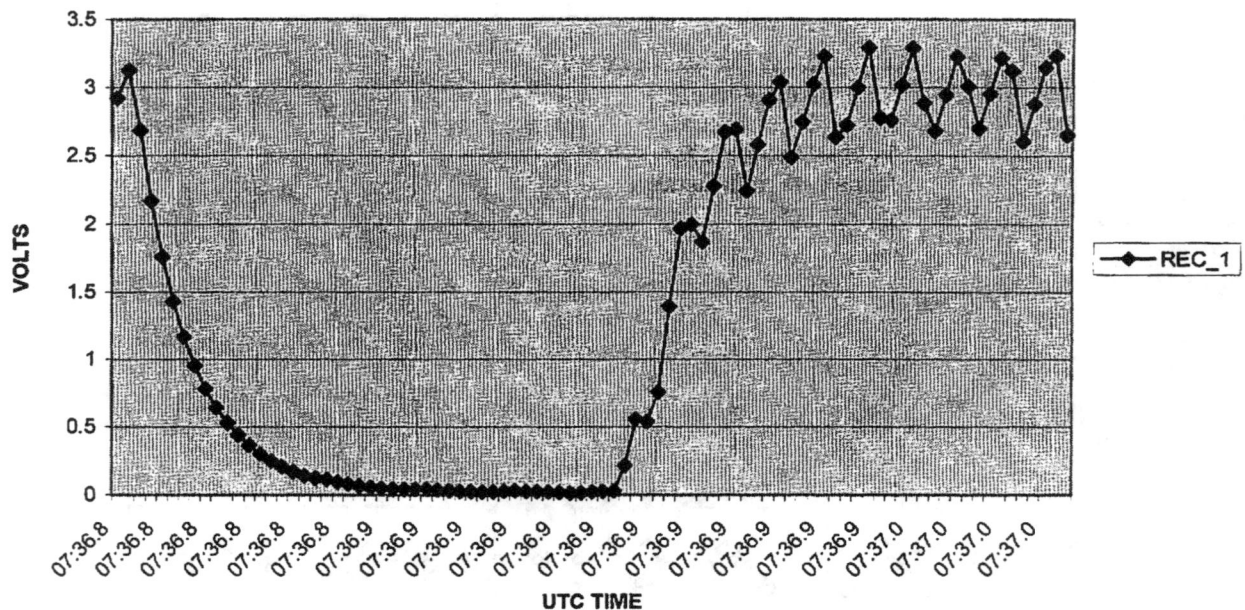

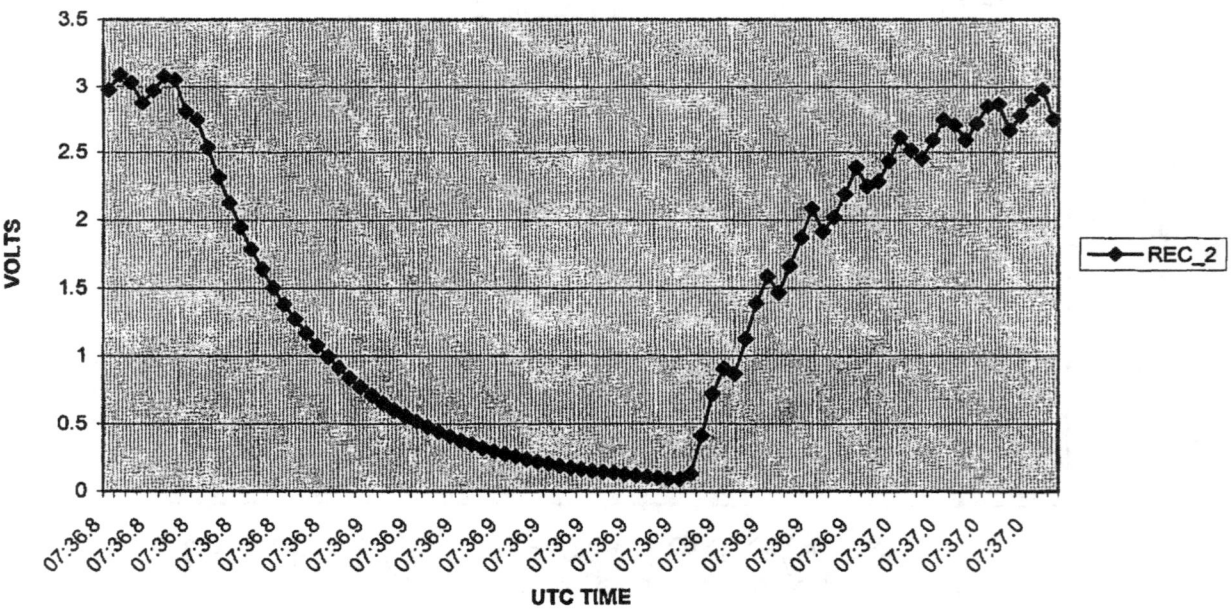

Figure 7. Excel Data Plots

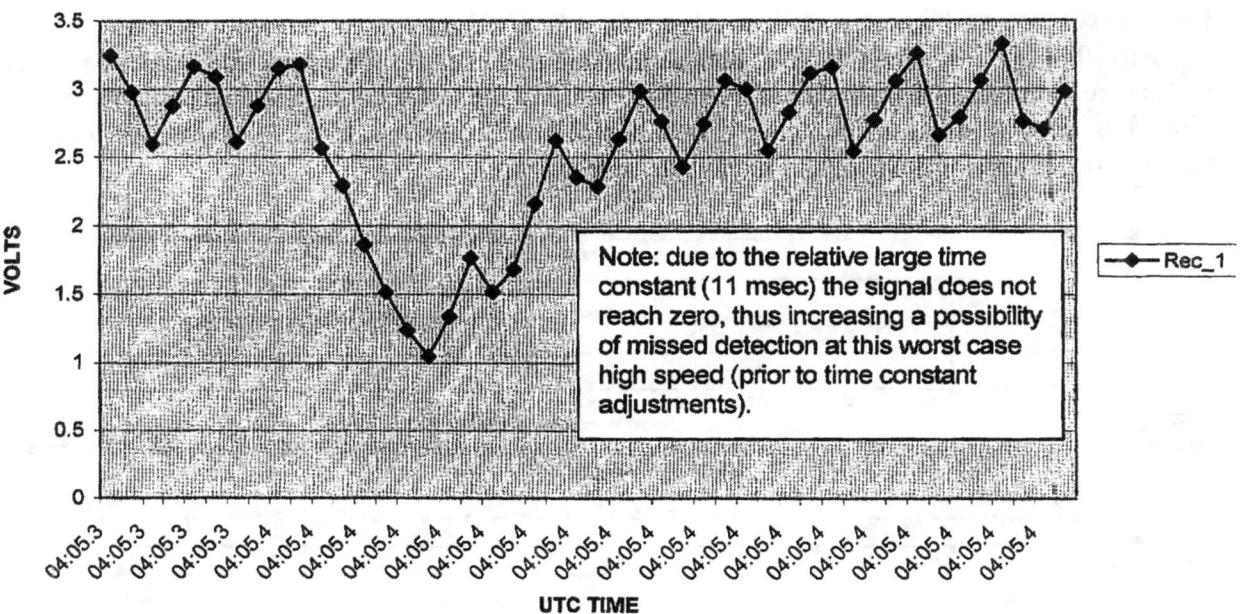

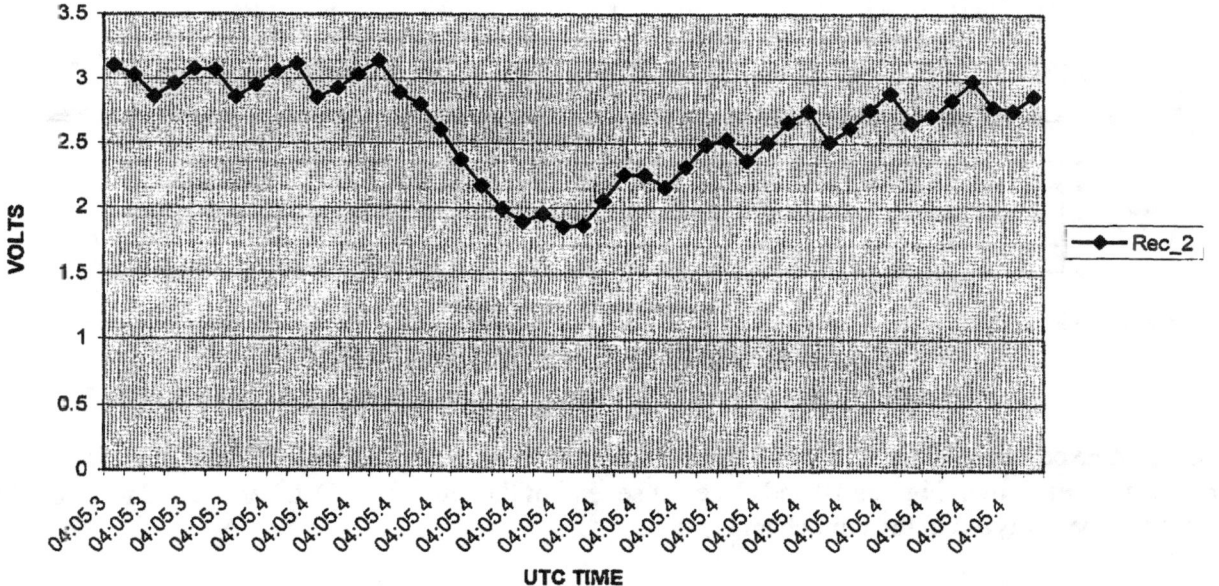

Figure 8. Excel Data Plots

4.1.2 Trip-Wire Test (Airfield – F-15s and Support Equipment)

The battery powered FogEye Systems were deployed to the taxiway Bravo on the airfield. The FogEye receivers each had 4 msec time constants. The UV light source frequency was changed from 60 hz to 1000 hz. The same CR-9000 data acquisition unit used for the compound test was utilized. Data acquisition was initiated just before an F-15 reached the test section. The data could then be plotted for analysis. Figure 9 depicts a CR-9000 plot from October 8, 2002. Note the two dips associated with the noise wheel and main gear, respectively.

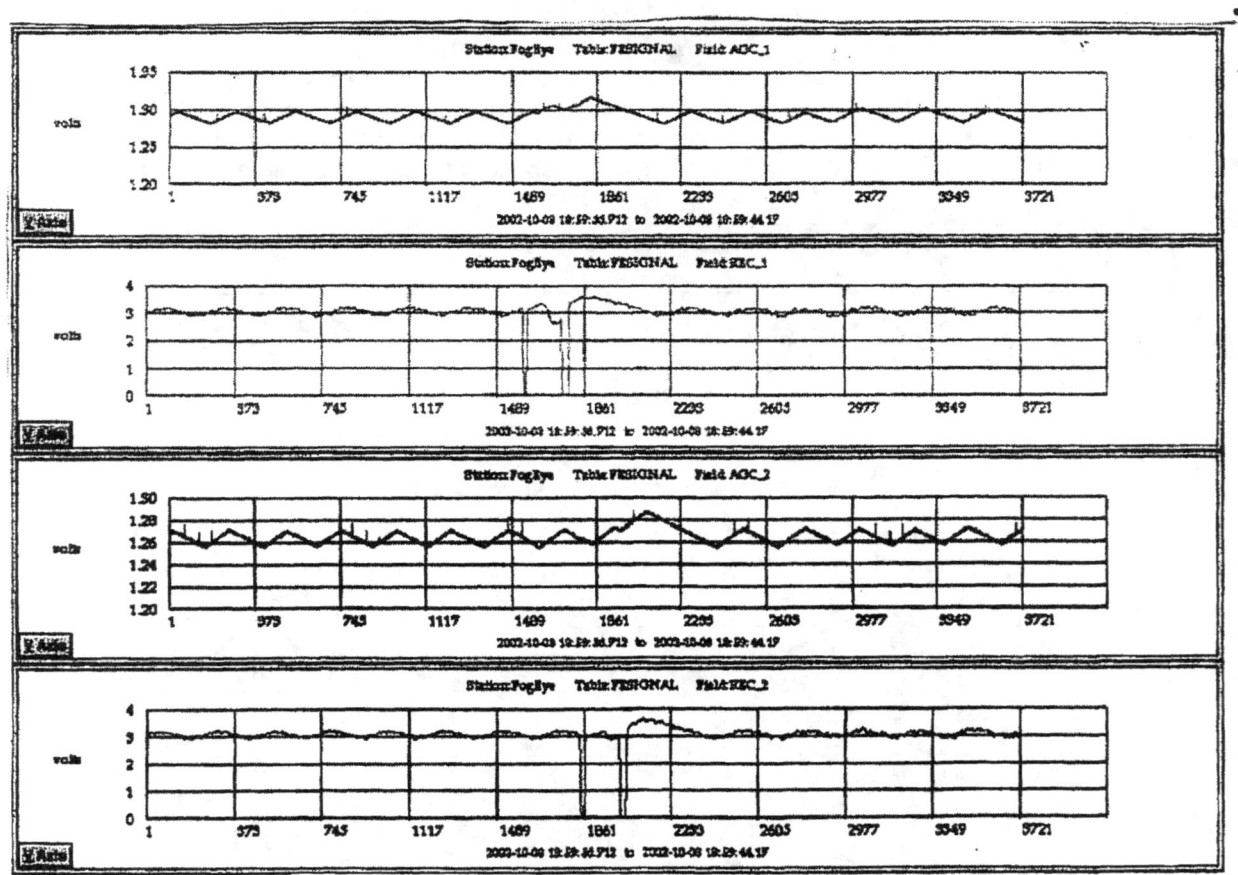

Figure 9. 10/08/02 Run 2 (30 mph) – F-15

Using an Excel spreadsheet, CR-9000 data was plotted on an expanded scale to detail the trigger and response of each run. Figures 10 and 11 depict several of the plots from October 8, 2002 testing. The plots will be analyzed in the next chapter.

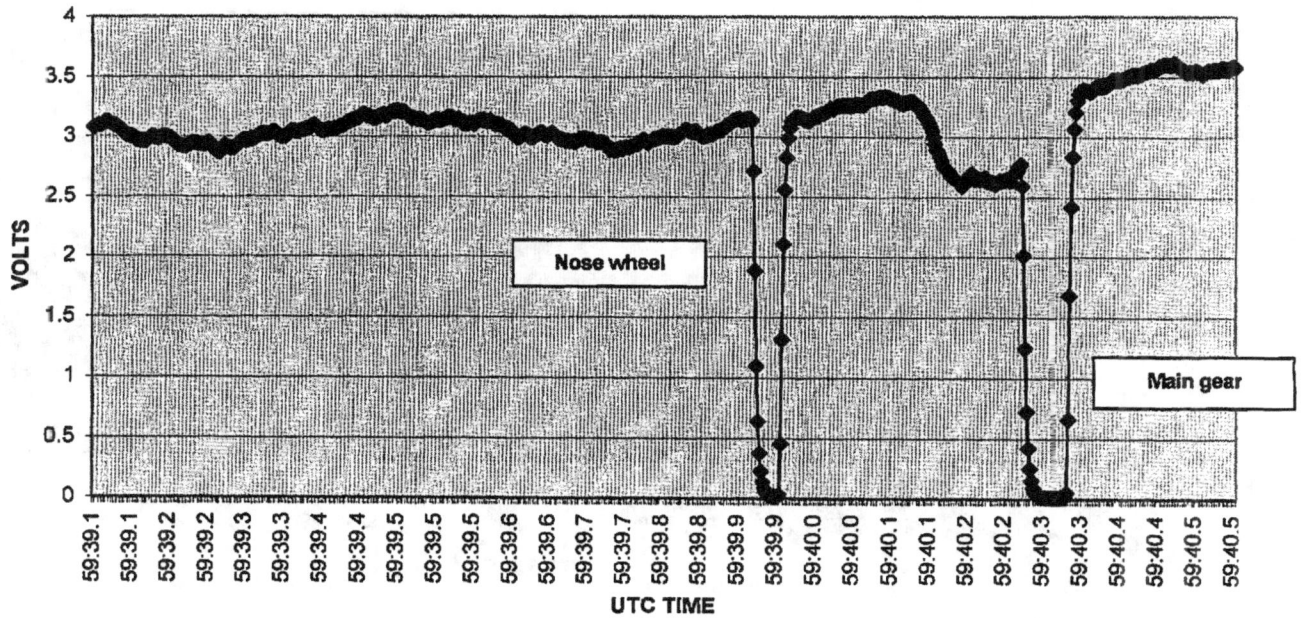

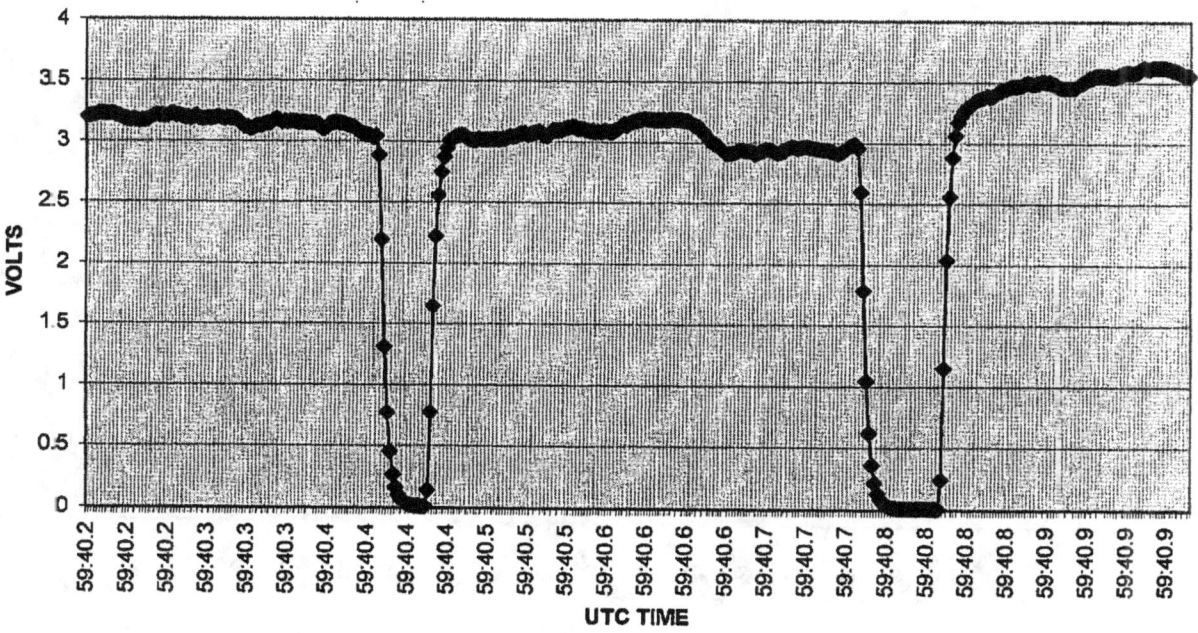

Figure 10. Excel Data Plots

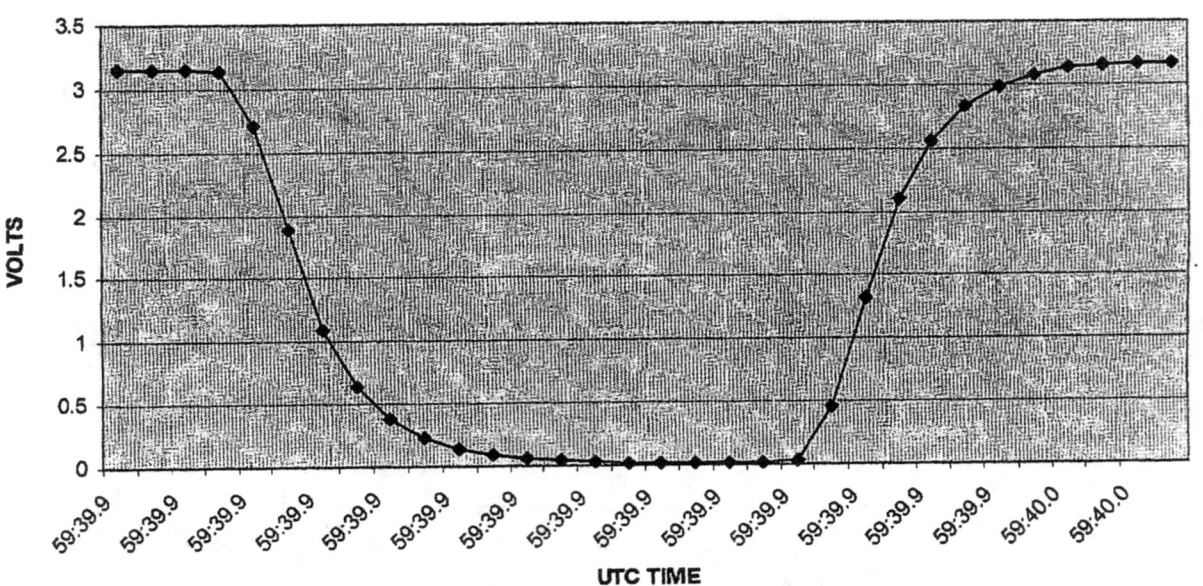

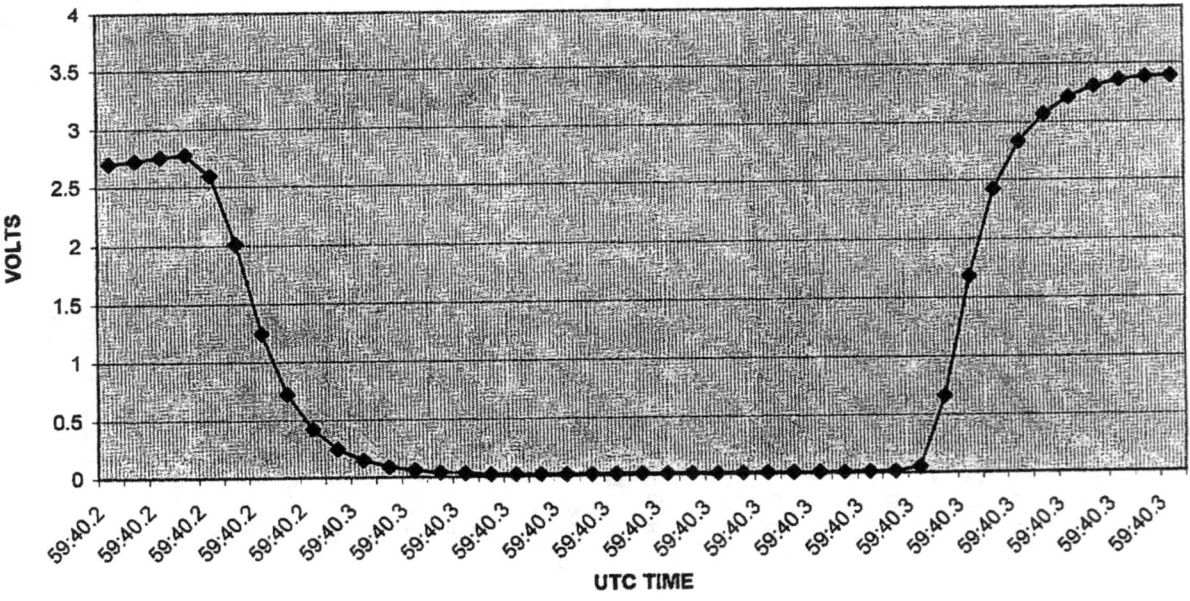

Figure 11. Excel Data Plots

4.1.3 Fog Test

Data recording for the 300-foot test used a Campbell Scientific 23X datalogger, which sampled at 10 Hz and averaged for 5 seconds. In addition to the 5-second averages, the standard deviations were calculated and recorded. The first sample of the new minute and the 11 prior samples from the prior minute were averaged to generate one-minute averages that were approximately synchronized with the Otis reference transmissometers.

4.2 AGC Response

Figure 12 shows the AGC response equations for the two trip-wire units. The manufacturer's calibration points for the individual detectors were used form a "best fit" curve that connected these points. The log-log form was selected because the interpolating lines between the three measurements are straighter and hence less likely to introduce errors. The curves were then used as transfer functions to transform the voltages into atmospheric transmission and scattering values.

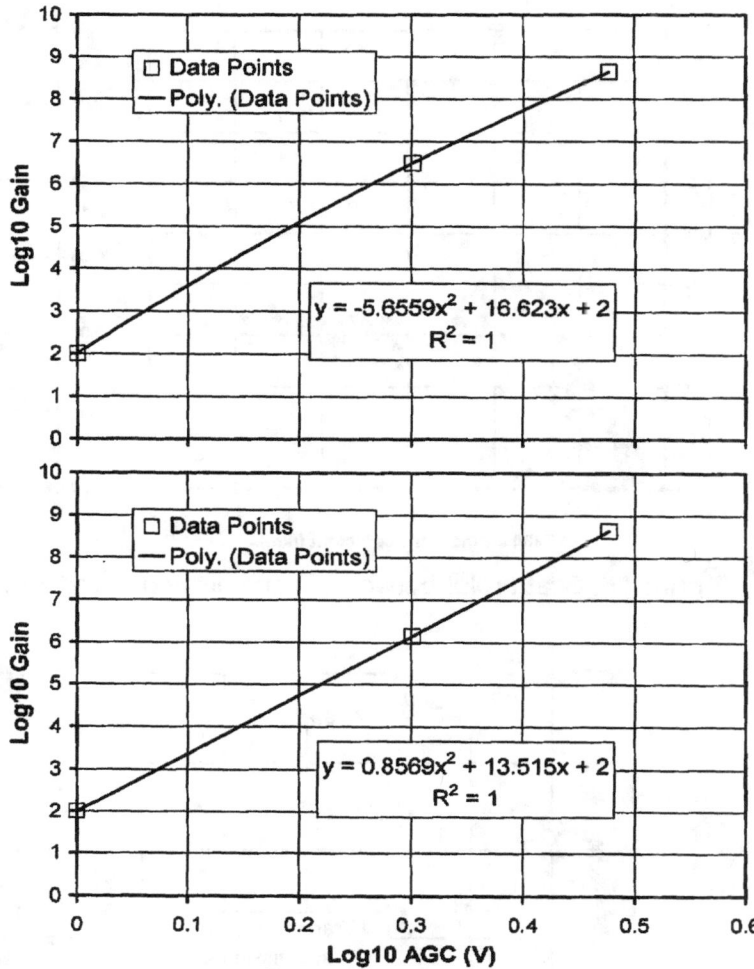

Figure 12. Receiver AGC Equations (Unit 2 top, Unit 3 bottom)

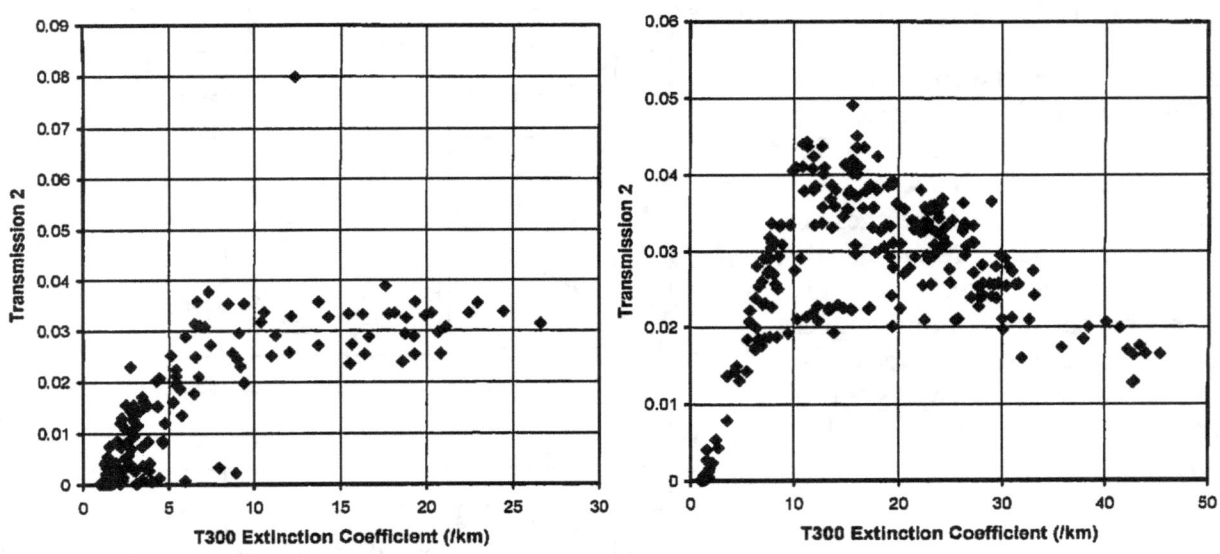

Figure 15. Scatterplot between Transmission 2 and Extinction Coefficient (9/19 left, 9/20 right)

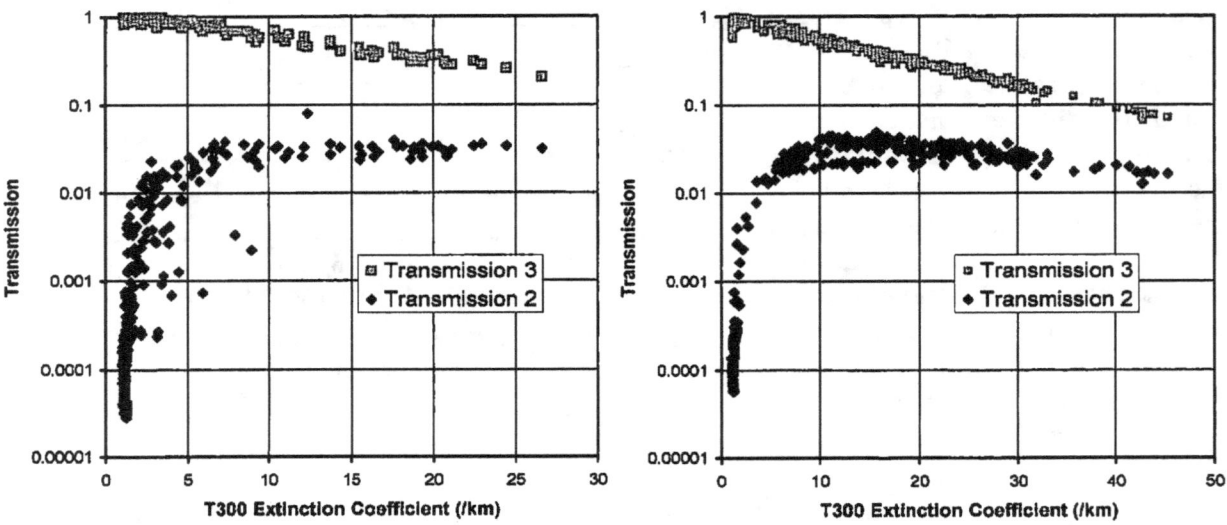

Figure 16. Trip-Wire Transmission vs. T300 Extinction Coefficient (9/19 left, 9/20 right)

is 300 feet and the meteorological optical range is 285 feet – all factors that border on worst case conditions.

An additional test was conducted at a later date to determine the impact of forward scattering on detection of beam interruption by a simulated 22.5-inch diameter wheel. These results shown in Figure 19, indicate the scattering is less, perhaps by a factor of 25%, than similar tests conducted with a 15-inch diameter wheel, as provided in Figure 15.

5.3.2 Transmissometer Unit 3

Figure 17 plots the extinction coefficient from FogEye Unit 3 for the same time period shown in Figure 14. The extinction coefficient is comparable to but smaller than the values from T300.

Figure 18 shows extinction coefficient scatterplots of FogEye Unit 3 versus T300 for 9/19/02 and 9/20/02. The fitted slope is about 70%, which suggests that some but not all of the forward scattered light remains within the 1.5-degree receiver beam. Note that the slope for 9/20 is reduced by high outliers at the minimum T300 extinction coefficient. The scatterplots[1] for the 15-degree Receiver 1 of FogEye Unit 1 had slopes closer to 50%.

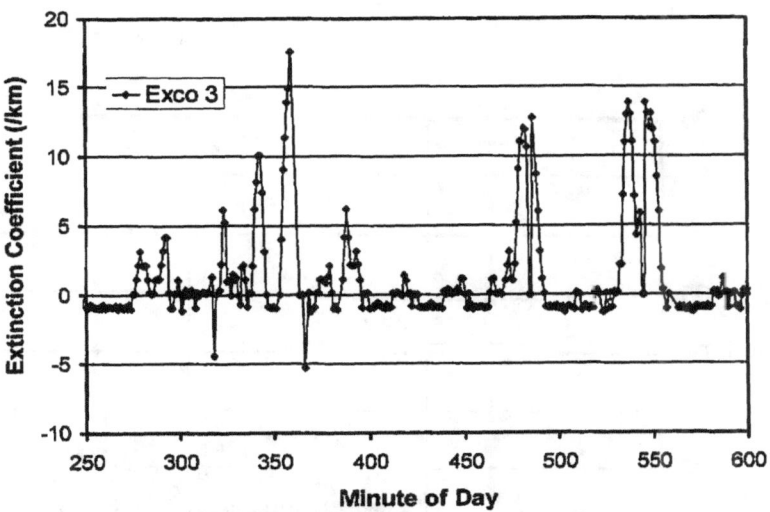

Figure 17. Extinction Coefficient from FogEye Unit 3 9/19/02

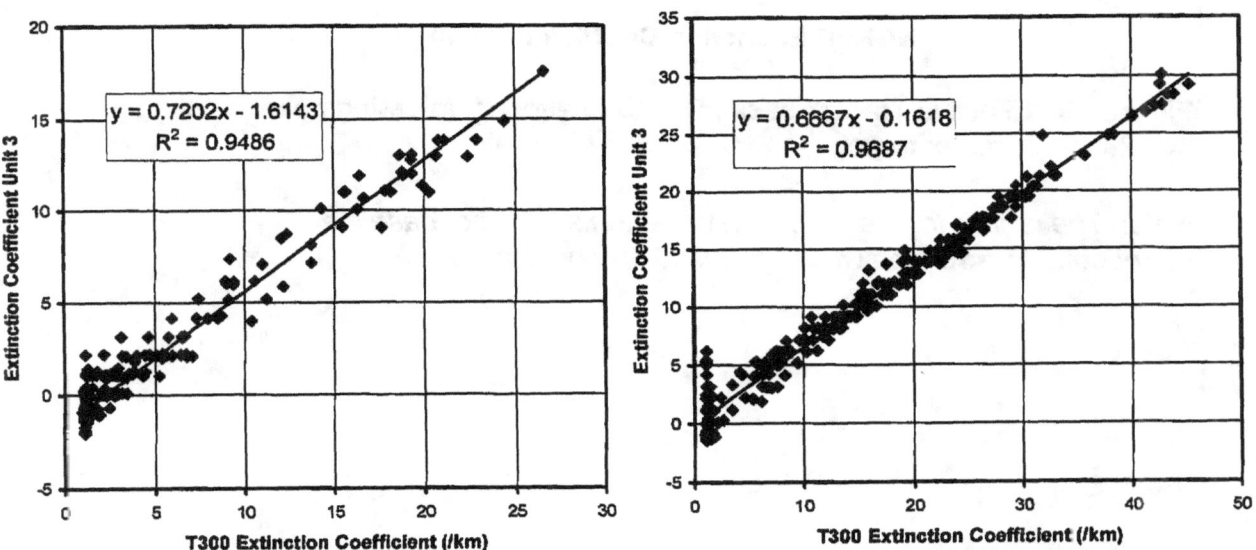

Figure 18. Extinction Coefficient Scatterplots: FogEye Unit 3 vs. T300 (9/19 left, 9/20 right)

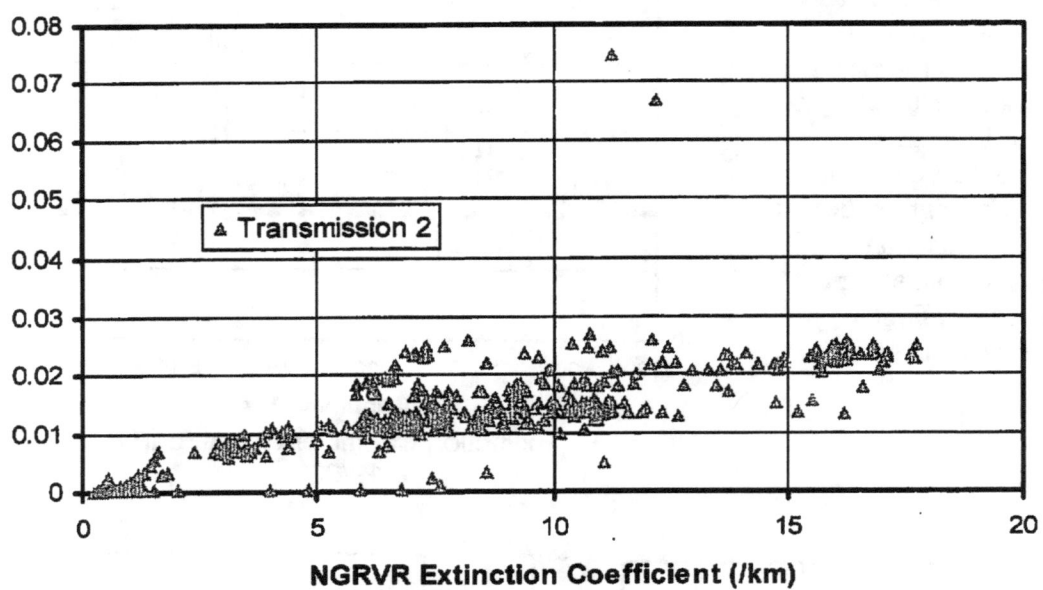

Figure 19. Effective Transmission, Due to Scattering, following beam blockage by 22-inch diameter target.

Note: The results for the 22.5-inch wheel on a 300-foot baseline correspond to a 15-inch wheel on a 200-foot baseline.

6. OPERATIONAL CONSIDERATIONS

6.1 FogEye Operational Capabilities

In 1994, the FAA published a table of user needs in recognition of a requirement for airport surface sensors to direct the surface movement of aircraft. This table related these needs to potential sensors that might employ different technologies to satisfy these needs. The table (please refer to Table 2) has been adapted to provide a convenient means to relate the capabilities of FogEye, a sensor technology not previously included in the table.

Table 2. User Needs vs. Sensor Type Matrix[2]

User Needs	FogEye UV	Inductive Loop	Pressure piezo electric	Microwave (Beam)	Infrared (Doppler)	Radar	Passive Mag.	Fiber Optic	Acoustic Active	Acoustic Passive	Optic
Stop Bar Control	Yes	Yes	Yes	Yes	Yes	Yes	Yes	Yes	Yes	Yes	Yes*
Taxiway Light Sequence	Yes	Yes	Yes	Yes	Yes	Yes	Yes	Yes	Yes	Yes	Yes*
"You are There/Memory	Yes	Yes[3]	Yes*	Yes	Yes	Yes	Yes	Yes	Yes*	X	Yes*
Runway Exit (clear of runway)	Yes	Yes*	Yes*	Yes*	Yes*	Yes*	Yes*	Yes*	Yes*	Yes*	Yes*
Direction	Yes	Yes*	Yes*	Yes	Yes	Yes	Yes	Yes	Yes*	Yes	Yes*
Movement Area Entry	Yes	Yes*	Yes*	Yes	Yes	Yes	Yes	Yes	Yes*	X	Yes*
Rate of Speed	Yes	Yes*	Yes*	Yes*	Yes	Yes*	Yes*	Yes*	Yes*	Yes*	Yes*
Aircraft/Vehicle Identification	No	No	No	No	No	No	No	No	No	No	Yes*
Aircraft/Vehicle Classification	No	Yes	Yes	No	Yes	No	Yes	No	Yes*	X	Yes*
Runway Status Lights	Yes	Yes*	Yes*	Yes*	Yes*	Yes*	Yes*	Yes*	Yes*	Yes*	Yes*
Flashing PAPI	Yes	Yes	Yes	No	Yes	No	Yes	No	Yes*	X	Yes*
Ground Marker	Yes	Yes*	Yes*	Yes*	Yes*	Yes*	Yes*	Yes*	Yes*	Yes*	Yes*

---

[2] RTCA/DO-221, "Guidance and Recommended Requirements for Airport Surface Movement Sensors, April 29, 1994
[3] Indicates some form of limitation

A FogEye signal, generated by an individual Transmitter-Receiver Trip Wire Sensor pair, can be used for automatic sequencing of stop bar and taxiway center line lead in lights (On-Off Switch), and for automated sequencing of segmented taxiway lights (Taxiway Light Sequencing), A combination of three pairs of these sensors can indicate the presence of an aircraft/vehicle at any given location on an airport (Presence/Memory), or give a positive indication of an aircraft/vehicle being clear of a runway (Runway Exit). The Presence/Memory and Runway Exit capabilities were among a number of capabilities demonstrated by Norris EO to FAA officials and others at Tipton Airfield on October 5, 2000. Their method of operation follows.

6.1.1   Presence/Memory

This capability can be illustrated by considering the use of three of these sensor pairs on a taxiway in a hold short area. The beam of one pair could be placed directly over the hold short line, nominally 225 feet from the runway centerline. A second beam could be placed between the hold short location and the runway, preferably 15 feet from the runway entrance. A third beam could be placed across the taxiway about 160 feet distant, and on the other side of the hold short line. The Receivers of each of these three pair would be interconnected such that the direction of travel of an aircraft on a particular taxiway would be known as well as its position relative to the hold short line, i.e., holding short, crossing the hold short line, entering the runway, or exiting the runway. An aircraft's presence is thus registered and maintained along a critical aircraft passage route. This architecture could be repeated for each taxiway and for other critical passage routes.

6.1.2   Runway Exit

This capability could be an extension of the Presence/Memory architecture. In this instance an aircraft is indicated as exiting the runway and being clear of same by tripping the third of the three beams that an aircraft encounters when exiting a runway. The lengths of the largest aircraft are about 230 feet. When the nose wheel of these aircraft intercepts the beam, the extreme of the aircraft's tail will be about 90 feet distant from the edge of the taxiway.

6.1.3   Discrimination Between Two Successive Aircraft

The FogEye Aircraft Presence Sensor must have the ability to distinguish between the passage of a single aircraft and the subsequent passage of a second aircraft that is immediately following. The pattern for each aircraft is (a) a beam interruption due to nose wheel, (b) non-beam interrupt period due to distance between the nose wheel and the main landing gear wheel(s), and (c) a second beam interrupt period due to the main landing gear wheel(s). The relative period lengths for this pattern, at a constant ground speed, are $a<c<b$. Satisfaction of this criteria indicates passage of an aircraft. The logic is then reset to begin this process again, after a time lapse of about b (minimum relative separation between aircraft), a value that may vary automatically from the equivalent of 20 feet minimum (separation between GA aircraft) to a maximum of 211 feet, separation from the tail of a "heavy" to the nose of a trailing aircraft.

6.2 FogEye Configuration Interface

Each Transmitter and Receiver that combine to function as a pair will be individually packaged as an integral part of existing runway or taxiway light fixtures. The packaging is such that an overall height of 14 inches is maintained for the augmented edge light fixtures. An interface diagram for operational units is provided in Figure 20.

6.3 FogEye Costs.

A preliminary budgetary recurring cost for each Transmitter and Receiver pair is $15,200, including installation. This is a total "on line" cost. The installation involves no surface penetrations. A complete hold short line installation, consisting of 3 sensor pairs and a control station is $53,000. The cost for a "guard" at each passage of a runway – runway intersection is $65,200. This installation is also ready to go on line. These costs are for initial units in quantities less than 5. Cost reductions of 20-40% for production prototypes and 20% for design simplification and manufacturing are anticipated.

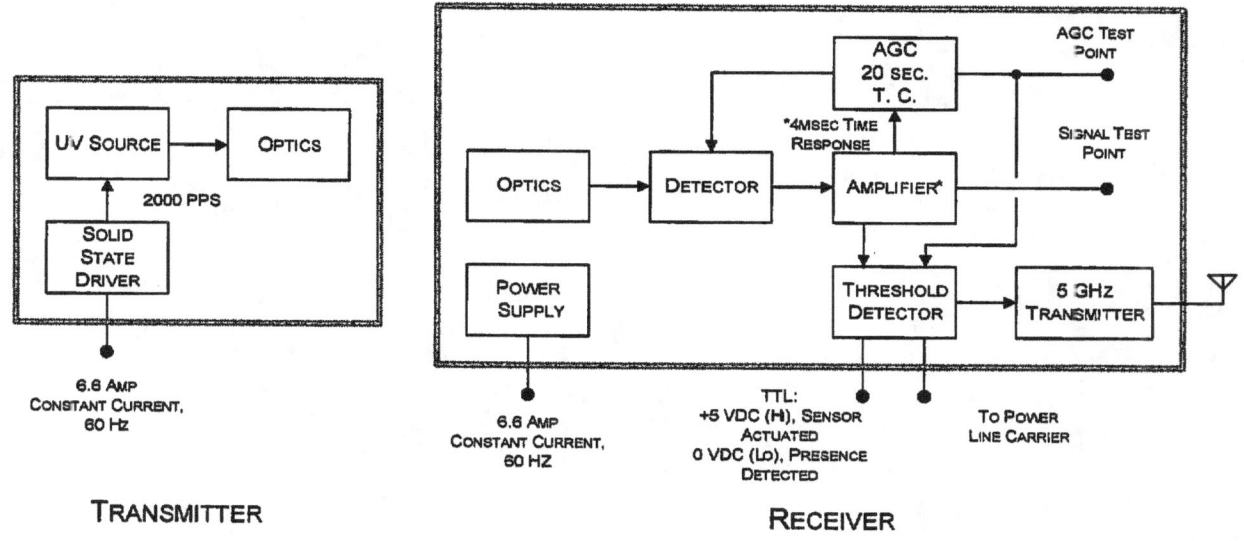

**INTERFACE BLOCK DIAGRAM
OPERATIONAL CONFIGURATION
FogEye™ AIRCRAFT PRESENCE SENSOR**

Figure 20

6.4 FogEye Reliability.

Transmitter and Receiver failures were experienced during the course of testing at the WTF. The Transmitter failures were due primarily to moisture induced breakdown of the wire insulation coating on inductors. The function of these inductors was subsequently replaced with solid state counterparts. The primary Receiver failures were traced to moisture induced high voltage leakage paths on PC boards. These failures were subsequently corrected with more effective conformal coatings. The electronic enclosures for both Transmitters and Receivers were exposed to extended driving rains and repeated solar heating cycles. They lacked adequate integrity to preclude water penetration and moisture accumulation. Subsequent configurations have addressed this issue.

7. CONCLUSIONS

1. FogEye Sensors tested were solar-blind.

2. The influence of forward scattering in fog on trip wire performance was studied on a 300-foot baseline. The drop in transmission caused by a 15-inch diameter wheel was found to be at least 7 db, which should be enough for reliable detection if a dynamic detection threshold is used to track the fog attenuation. Better performance would be expected for shorter baselines or larger wheel diameters. Also, the existence of two distinct signal channels, each with its own time response can be incorporated, with knowledge of the signal dynamics in fog, to produce a third channel. This channel could provide automatic threshold adjustment and thereby a performance margin that may be considerably greater than 7 db.

3. The AC-powered FogEye trip-wire sensors performed within expectations as a trip-wire system for detection of aircraft and vehicle movement, but their dynamic range was limited by 120-hz transmitter ripple and long time constants. The limited dynamic range would compromise performance when the signals are reduced by fog attenuation.

4. The battery-powered FogEye trip-wire sensors used 1000-hz lamp excitation and 4-msec time constants and were found to have a full dynamic range.

5. Operational considerations indicate that as a primary component of a sensor subsystem the FogEye sensor would be well suited for use with stop bar, runway status lights, and taxiway light sequencing. Referring back to Table 1, FogEye would have a yes in every category except aircraft/vehicle identification and aircraft/vehicle classification.

8. FUTURE TESTING

Phase-III Measurement of the UV emissions from existing aircraft lights will take place at BWI Airport. The objective of these measurements is to determine the ability of the FogEye System to detect the presence of non-modified aircraft during three different phases; short final approach; landing and rollout; position and hold; and takeoff. .

Tests of the FogEye Trip Wires on the 300 foot baseline (para. 3.2) will continue with larger wheel diameters to provide additional data on the extent of trip-wire dynamic range in fog.

www.ingramcontent.com/pod-product-compliance
Lightning Source LLC
Chambersburg PA
CBHW081809170526
45167CB00008B/3383